D1784880

Census 2001

Key Statistics for postcode sectors in England and Wales

Laid before Parliament pursuant to Section 4(1) Census Act 1920

Office for National Statistics

London: TSO

30109 02161989 8

© Crown copyright 2004.
Published with the permission of the Controller of Her Majesty's Stationery Office (HMSO)

ISBN 0 11 621749 9

This publication, excluding logos, may be reproduced free of charge in any format or medium for research or private study subject to it being reproduced accurately and not used in a misleading context. The material must be acknowledged as crown copyright and the title of the publication specified. This publication can also be accessed at the National Statistics website: **www.statistics.gov.uk**

For any other use of this material please apply for a free Click-Use Licence on the HMSO website:
www.hmso.gov.uk/click-use-home.htm
or write to HMSO at The Licensing Division, St Clements House, 2-16 Colegate, Norwich NR3 1BQ Fax: 01603 723000 or e-mail:
hmsolicensing@cabinetoffice.x.gsi.gov.uk

Contact points

For enquiries about the 2001 Census or this publication, contact the National Statistics Customer Enquiry Centre on
0845 601 3034
(minicom: 01633 812399)
E-mail: **info@statistics.gov.uk**
Fax: 01633 652747
Post: Room D115,
Government Buildings,
Cardiff Road,
Newport NP10 8XG

To order this publication, call The Stationery Office on **0870 600 5522**. See also back cover.

About the Office for National Statistics

The Office for National Statistics (ONS) is the government agency responsible for compiling, analysing and disseminating many of the United Kingdom's economic, social and demographic statistics, including the retail prices index, trade figures and labour market data, as well as the periodic census of the population and health statistics. It is also the agency that administers the statutory registration of births, marriages and deaths. The Director of ONS is the National Statistician and the Registrar General for England and Wales.

A National Statistics publication

National Statistics are produced to high professional standards set out in the National Statistics Code of Practice. They undergo regular quality assurance reviews to ensure that they meet customer needs. They are produced free from any political interference.

BEXLEY LIBRARY SERVICE

LOC CR | CL NO. 384 60942
PRICE | ACC DATE MAY 04 | BKS PA
LL CODE REF | ITEM LOAN TYPE REF
| LANG | CH | PR

Contents

Foreword

Results from the Census are made possible by the co-operation of the public in responding to the Census; by the hard work of the Census field-staff; and by the assistance of many other people and organisations that have supported all aspects of the Census. The Registrar General would like to thank all those who have contributed to the Census.

Len Cook

Registrar General for England and Wales

Introduction

This report with the accompanying CD provides results from the 2001 Census for postcode areas, districts and sectors throughout England and Wales, and is published under the authority of the Census Act 1920.

The statistical results, which cover all topics in the Census, are on the CD. Outlines of the tables are in *Appendix C*, and help using the CD is in *Appendix B*. The printed part of the report has background information on postal areas, and on the 2001 Census in general.

Postal areas

Royal Mail maintains a system of freely available postcodes for every address in the UK for the handling and delivery of mail which has also become a common basis for organising other information collected in the public and private sectors. Census results for postal areas are provided to support this wider use.

Postcodes have an 'outward' part of two to four alphanumerical characters, and an 'inward' part of three alphanumerical characters, for example

PO16	7DZ
outward	inward
code	code

They have four hierarchical levels

Example	Geographic unit	Number in UK (August 2003)
PO	Postcode area	124
PO16	Postcode district	2,932
PO16 7	Postcode sector	9,750
PO16 7DZ	Unit postcode	1.75 million (approx)

The 1.75 million unit postcodes cover 27 million delivery points, 0.22 million of which are 'large users' with large volumes of mail.

Fitting Census information to postal areas

The results in this report provide the best possible fit to the populations of postcode areas, districts and sectors at the time of the Census in April 2001, but the definitions of these areas may differ from other sources.

Postal areas have no definitive boundaries, and the codes refer to groups of addresses. The statistics in this report represent residents and households at addresses in a particular postal area. The 2001 census returns were postcoded, but the codes were not used directly to produce the statistics for this report for reasons explained below in the section on Confidentiality.

Instead, postcode units with residents were grouped to form 'Output Areas', each with around 125 households - the smallest areas for which detailed Census results are presented and which form bricks to build to higher geographies such as postal areas (a description of the creation of Output Areas is provided at www.statistics.gov.uk/census2001/op12.asp). However, Output Area boundaries follow administrative boundaries – parishes, wards and local authorities– whereas postal areas do not, and the addresses in a postal area may sometimes be split between Output Areas.

In consequence, the Output Area building bricks have been 'best fitted' to postal areas where all addresses in an Output Area do not fall into a single postal area, on the basis of the postcodes of the majority of the population. When this happens, the Census returns for one postal area may be included in the statistics for another, as each Output Area is allocated as a whole to a postal area. However, the 'best fitting' is at the margins of the relatively large postcode sectors, or the larger postcode districts and areas, and tends to be self-cancelling.

There have also been changes in postal areas since April 2001. The system frequently changes to accommodate new addresses or changes to and from 'large user' status. Whole areas may be re-coded, and expired codes may be re-used in different locations. Details of larger reorganisations are provided in the Royal Mail 'Postcode Update' series, and the most recent, together with a summary of major changes since 1990, are available at www.royalmail.com

Confidentiality

The Key Statistics and Profiles featured in this report are not released for areas with fewer than 100 residents or 40 households as one of a number of measures to prevent the inadvertent disclosure of Census information about identifiable individuals. Postal areas with populations which fall below either of these thresholds are either subsumed in a nearby area through the process of 'best fitting' Output Areas or less often gain sufficient population from other areas to be above the threshold and to be included in the report.

The 'best fitting' of the Output Area building bricks to postal areas is also another measure taken to stop inadvertent disclosure as it prevents the possible risk from the differencing of standard sets of statistics for overlapping areas, such as a postcode sector and a ward, which might isolate Census information for small populations.

Locating postal areas

The user of this report will need to have some external information to locate postal areas as maps are not included. Available maps of postal areas can provide a general location, but the boundaries will not be defined around the residential addresses in each postal area which are the basis for the results in this report, and the maps may not show the areas at the time of the Census in April 2001.

Boundaries for postal areas presented in this report may be created by using the vector (re-usable) boundaries for Output Areas and the look-up table of postal areas by Output Areas, both available on request from Census Customer Services (*see page 5*), and a Geographical Information System to select and reproduce those boundaries which relate to postal areas. The Output Areas boundaries are supplied under terms and conditions which reflect certain Ordnance Survey rights which may apply for use outside an organisation's own business.

Census results for postal areas

On the CD accompanying this Report

Key Statistics

Key Statistics provide information for all topics covered by the Census. There are approximately thirty Key Statistics tables and each is designed to summarise information relating to a single topic. A list and description of the tables is included in *Appendix C*.

Postcode Sector Profiles

The Postcode Sector Profiles are designed to provide briefer summary results. There are three profiles, relating to 'All people resident in the area', 'All people aged 16 – 74 resident in the area', and to 'Households resident in the area'. Descriptions of the profiles are included in *Appendix C*

On request

Census Area Statistics (CAS)

This set of tables is released for Census Output Areas and for areas built from them, and provides more detailed results than the Key Statistics, particularly through cross-tabulations showing relationships between two or more topics. There are also tables which provide more detailed on a single topic.

CAS may be aggregated to postal areas, or to other higher geographies, from the CAS for Output Areas with the use of look-up tables. These and CAS are available from Census Customer Services (*see page 5*). Alternatively, Customer Services may be contacted for information about the availability of CAS produced for postal areas.

Finding information in this Report

The main screen shows a grid with an expandable list of areas down the left-hand side, and a list of tables across the top. The expandable list allows you to view results for named postcode areas and their postcode districts. The postcode areas in the list are split into ten groups based on their postcode area code (rather than the postcode area name). To view results for EC1 (London East Central), for example, you would click on the '+' next to the E – F group to view a list of the postcode area codes starting with those letters. Then, click on the icon in the row labelled EC and in the column relating to the table of interest. You will then be shown Census results on that topic for postcode areas and postcode districts.

When viewing these results, you can also choose to view results for these areas and postcode sectors as counts or percentages using SuperTABLE by clicking on the 'click here to download or manipulate data' icon above the results. You will be presented with options to select either a file of percentages or a file containing counts. Both these options enable you to access results for all postcode areas, districts and sectors in England and Wales in one SuperTABLE file. On selecting either of these options you will be taken to the top of a SuperTABLE file holding results for all postcode areas, districts and sectors, and you will then be able to scroll down the table to find the area of interest within the file. Again, areas are arranged in alphanumeric order of the postal code, rather than postal name. SuperTABLE also allows you to view and manipulate the postcode sector profile (the first table listed across the top of the grid) which provides some key results on a range of topics for that area. You will need to have installed software available on this and previous 2001 Census CDs to view the tables in this format.

If you know the name of the postal area or district, but don't know the code, you can use the 'Search for areas' box above the grid on the main screen to find the area you need.

Postcode sectors with very small residential populations will not be shown separately in the tables. Counts for these areas appear in the results for neighbouring areas.

Explanations of terms used in tables are provided in the *Glossary*, and in the *Classifications*. Where a particular explanation is required within a table, this is indicated by a footnote marker and the information provided in a footnote to the table.

More detailed information on the terms used in tables; comparability with the 1991 Census; the Census questions; and response and imputation rates will made available on the National Statistics website and published in other reports, as described in *Further results from the 2001 Census*.

Populations covered in this Report

Each table on the CD relates to a group of people or households, referred to as the 'table population'. Many tables relate to the population 'All people', but other tables relate to subsets of 'All people', such as 'People in employment aged 16 – 74', or to distinct populations such as 'All households'

People

In the 2001 Census information was collected only on usual residents. This contrasts to the 1991 Census which collected information on both usual residents and visitors on Census night. A usual resident is generally defined as someone who spends most of their time residing at that address. It includes:

- People who usually live at that address but are temporarily away from home (on holiday, visiting friends or relatives, or temporarily in a hospital or similar establishment) on Census Day.
- People who work away from home for part of the time, or are members of the Armed Forces.
- Students if it is their term-time address
- A baby born before 30th April even if it is still in hospital.
- People present on Census Day, even if temporarily, who have no other usual address.

However, it does not include:

- Anyone present on Census Day who has another usual address.
- Anyone who has been living or intends to live in a special establishment such as a residential home, nursing home or hospital for six months or more.

Households

A household is defined as one person living alone, or a group of people (not necessarily related) living at the same address with common housekeeping - that is, sharing either a living room or sitting room or at least one meal a day.

Communal establishments

A communal establishment is defined as an establishment providing managed residential accommodation. 'Managed' means full-time or part-time supervision of the accommodation.

In most cases (for example, prisons, large hospitals, hotels) communal establishments can be easily identified. Identifications is less easy with small hotels, guest houses and sheltered accommodation. Special rules apply in these cases:

Small hotels and guest houses are treated as communal establishments if they have the capacity to have 10 or more guests, excluding the owner/manager and his/her family.

Sheltered housing is treated as a communal establishment if less than half the residents possess their own facilities for cooking. If half or more possess their own facilities for cooking (regardless of use) the whole establishment is treated as separate households.

More information on the definition of table populations can be found in the *Glossary* contained on the CD.

Quality of the results

The use of the One Number Census methodology (see *Appendix A: Background information on the 2001 Census*) means that the results of the 2001 Census cover the entire population of England and Wales, and are the most reliable achievable. However, there are a number of sources of potential error in the results. These include

- 'Incorrect' information provided on the forms.
- Sampling error related to estimates derived through the One Number Census process.
- Errors introduced during processing (such as coding errors).

Some elements of 'incorrect' information will have been corrected during the edit process (see *Appendix A: Background information on the 2001 Census*). Following this, the results

have undergone an extensive quality assurance process, including checks against administrative records and sources of information on particular groups such as students and the armed forces.

As the Census results, which incorporate an adjustment for under-enumeration through the One Number Census methodology, are estimates based partly on a sample survey, sampling errors can be used as a guide in assessing the accuracy of the results. The sampling error can be used to construct a 95 per cent confidence interval - that is a range in which we can be 95 per cent confident that the true value lies. For the population of England and Wales, this confidence interval is 52,041,916 +/-104,000 (0.2 per cent of the estimated population). An important aspect of the One Number Census methodology is that the estimates are unbiased - that is, that they are not systematically above or below the true value.

Further information on coverage and imputation rates to individual questions is available at www.statistics.gov.uk/census2001/methodology.asp. This material now includes quality indicators for small areas in the form of a rating scale based on the One Number Census imputation rates, as well as overall response rates and estimates of sampling errors for Local Authority Districts. A detailed report covering a range of aspects of quality of the Census results will be published in 2004.

Comparison of the results with those from the 1991 Census

Comparison of the results in this Report with those of the 1991 Census must be treated with caution as the comparison of results from the 1991 and 2001 Censuses is affected by three factors.

Changes in definition

There are a number of differences in definitions and information collected between the 1991 and 2001 Census. More information on this will be made available in the 2001 Census Definitions volume and on the National Statistics website.

Changes in the geographic base

As previously mentioned, changes in geographic boundaries between 1991 and 2001 may mean that results which overtly relate to the same named area actually relate to different boundaries.

Adjustment for under-enumeration

Results of the 2001 Census have been adjusted, via the One Number Census process, to account for under-enumeration. As results of the 1991 Census were not subject to the same methodology, direct comparisons with the 1991 results must be undertaken with caution.

Where comparisons between the 1991 and 2001 Census results are required, the effects of the above changes are in most cases minimal but can be further mitigated by comparing differences between percentages calculated from the respective bases in each census rather than measuring the percentage difference between the actual counts at each Census.

Other Censuses in the UK

Separate Censuses were carried out, on the same day and using similar methodologies, in Scotland and Northern Ireland, under the authority of the respective devolved administration and Registrars General. Information on these Censuses is available from:

Scotland

General Register Office for Scotland
Statistics Division
Ladywell House
Ladywell Road
Edinburgh
EH12 7TF
Tel: 0131 314 4254
E-mail: customer@gro-scotland.gov.uk

Northern Ireland

Census Customer Services
Northern Ireland Statistics and Research Agency
McAuley House
2-14 Castle Street
Belfast
BT1 1SA
Tel: (028) 9034 8160
Fax: (028) 9034 8161
Email: census.nisra@dfpni.gov.uk

Further results from the 2001 Census

Further results from the 2001 Census provide information for a range of administrative areas and other geographies in common use. In addition to the statistical reports, supporting information on the Census, including definitions of Census terms; an evaluation of the quality

of the Census results; and information on geographies used in the Census results will also be published. Further information on results from the 2001 Census, including information on commissioning output, is available on the National Statistics website www.statistics.gov.uk/census2001/op.asp or from Census Customer Services census.customerservices@ons.gov.uk

Copyright and reproduction of material from this Report

This publication (excluding the departmental logo) may be reproduced free of charge in any format or medium for research or private study. This is subject to it being reproduced accurately and not used in a misleading context. The material must be acknowledged as Crown copyright and the title of the publication specified.

Material in this publication can also be accessed at the National Statistics website at www.statistics.gov.uk. For any other use of this material please apply for a free Core Click-Use Licence on the HMSO website at www.hmso.gov.uk/click-use-home.htm, or by writing to HMSO at

The Licensing Division
St Clements House
2-16 Colegate
Norwich NR3 1BQ
Fax: +44 (0)1603 723000
E-mail: hmsolicensing@cabinetoffice.x.gsi.gov.uk

Further information

Further information on the 2001 Census is available on the National Statistics website at www.statistics.gov.uk or from Census Customer Services at:

Census Customer Services
ONS
Titchfield
Fareham
Hants PO15 5RR
Telephone: ++44 (0) 1329 813800
Fax: ++44(0)1329 813587
Minicom: ++44(0)1329 813669
E-mail: census.customerservices@ons.gov.uk

Notes to tables

Adjustment of small counts

Note that small counts in tables have been adjusted to prevent the disclosure of information about identifiable individuals. This means that different tables may show different counts of the same population.

Percentages

Percentages and derived statistics are shown to two decimal places.

Small postcode sectors

Postcode sectors with very small residential populations will not be shown separately in tables. Counts for these areas appear in the results for neighbouring areas.

Table KS01

Information on the population in 1991 which was provided for local authorities in Table KS01 in a previous product, is not available for postcode sectors, and these columns have been left blank in this table on the CD.

Table KS02

Mean age is calculated as the mean of ages at last birthday. Mean elapsed age, taking into account that fraction of the year between birthday and Census Day can be approximated by adding 0.5 to the shown figure.

Table KS06

Two versions of KS06 (Ethnicity) are provided on the CD. The first provides results for the standard ethnic groupings for postcode areas, districts and sectors throughout England and Wales. The second version provides, in addition, information on identification as Welsh for postcode sectors which fall wholly in Wales. Information is not provided for postcode areas or districts. Corresponding information for Wales as a whole is available in, for example, Key Statistics for local authorities in England and Wales. Information on identification as Welsh within the Welsh part of postcode sectors straddling the England-Wales border can be obtained using the Key Statistics for Output Areas, in conjunction with the Output Area to higher areas lookup file, or by contacting Census Customer Services.

Table KS25

Table KS25 (Welsh Language) is produced only for postcode sectors which fall wholly in Wales. Information is not provided for postcode areas or districts. Corresponding information on skills in the Welsh language within the Welsh part of postcode sectors straddling the England-Wales border can be obtained using the Key Statistics for Output Areas, in conjunction with the Output Area to higher areas lookup file, or by contacting Census Customer Services.

Appendix A: Background information on the 2001 Census

Legislation

The 1920 Census Act allows for the carrying out of a census no less than five years after the previous census. However, various other legislative requirements need to be fulfilled before a census can be held. The first stage in the 2001 process was the publication of the White Paper *The 2001 Census of Population*[1] which sets out the reasons for holding a census, the proposed questions, operational methodology and format of results. The White Paper was produced in March 1999, two years before the Census, to ensure sufficient time for public discussion of proposals.

Prior to every Census, a Census Order stating the date of the Census, the people who are required to complete the form, those who are to be included on the form and the topics on which questions will be asked is required to be approved by Parliament. The operational aspects of the Census also required Parliamentary approval and this information was set out in the Census Regulations, which contained details of how the Census was to be conducted and included a copy of the census form.

The decision to include a question on religion in the 2001 Census, resulted in additional legislation, which required amendments to be made to the 1920 Census Act, Census Order 2000 and Census Regulations 2000 to make provision for this new question to be asked.

Fieldwork

The Census was designed to collect information on the resident population on Census Day - 29 April 2001. Before this day, enumerators were employed to deliver Census forms to every identified household space and communal establishment. Residents were asked to complete the forms with the information as correct on Census Day, and to post the completed forms back in a pre-addressed envelope. Where a form was not received through the post after a specified period, the enumerator visited the address in order to collect the form by hand. Special arrangements were made to enumerate the Armed Forces and people sleeping rough. The overall response rate (that is, the proportion of people included on a returned Census form)

was estimated to be 94 per cent. Information on a further 4 per cent of the population was collected by enumerators.

Households absent from their usual address on Census day were required to complete a form on their return to that address (though many had done so beforehand). Where they did not, and in other instances where a household failed to return a form, the enumerator recorded the type of accommodation and an estimate of the number of rooms and number of residents. This information was used within the One Number Census process in adjusting the results for underenumeration in the Census.

The Census was followed by the Census Coverage Survey (CCS), which took place between 24 May and 18 June 2001. This was an independent doorstep survey of a sample of a third of a million households, covering every local authority, which was used to adjust the Census counts for under-enumeration (see *The One Number Census* below). The Census Coverage Survey returned a household response rate (the proportion of identified households which were successfully interviewed) of 91 per cent.

Processing

Returned forms were fed through scanning machinery which captured all the ticked responses and stored written answers in digital form. The latter were coded into categories either by automatic systems that recognise terms given in response to questions, or by manual coding. This data then underwent an edit process to ensure that the data was consistent, and an imputation process to estimate responses for questions which had not been completed on the original form.

The One Number Census

A key strategic development for the 2001 Census was the adoption of the One Number Census process. This was used to adjust the results of the 2001 Census to take account of the fact that a single attempt at counting the population (the Census) never counts everyone. The results of the Census were matched against those of the Census Coverage Survey. This enabled the numbers and characteristics of the total population to be estimated, including those not counted by either the Census or CCS. Data from administrative registers and demographic estimates were used to quality assure these estimates. All results from the 2001 Census thus incorporate allowances for Census underenumeration. Further information on the methodology of the One Number Census is available on the National Statistics website.

[1] *The 2001 Census of Population, CM4253. HMSO, 1999.*
  ISBN 0-10-142532-5

Appendix B: Using the CD

The CD accompanying this Report will work on PCs with the recommended minimum system requirements:

CPU 200MMX
64MB RAM
4x CD-ROM Drive
400MB free drive space
1024 x 768 screen resolution*
Internet Explorer 5, Netscape 6, and above
    Browsers
For Windows NT, Service Pack 6 is required
    as a minimum.

* Check your screen resolution is set to at least 1024 x 768. To do this, for PC's, select My Computer from your Desktop, then Control Panel, then Display. Select Settings and adjust your screen area with the control provided.

If your PC meets the above requirements you can view the Key Statistics for postal areas and districts in the Browser. If you wish to view the Key Statistics or profiles for postcode sectors you will need to install some additional software.

Insert the CD into the CD drive and you will be presented with the terms and conditions of supply of the Census results and an invitation to view a Read Me file. Once you have read this, and minimised or closed that file, you will be asked to accept the terms and conditions. If you accept, you will be presented with a screen allowing you to view tables for postal areas and districts, or to install the additional software: SuperTABLE which is needed to view the Key Statistics and profile tables for postcode sectors; and Adobe Acrobat which is needed to view the example postal geography maps.

More information on using the CD and installing the software is provided on the CD itself, in the Read Me file shown when first using the CD.

Apple Mac users will be able to view results for postcode areas and districts by following the above instructions, but will not be able to install the software necessary to view results for postcode sectors, or to download and manipulate the results for postcode areas and districts.

Appendix C: Tables included on the CD

Table **KS01**

Usual resident population

All people

Postcode sectors in England and Wales

Area	1991 Resident population[1]			2001 Population:			Percentage intercensal population change[1,2]	Percentage of people living in households	Percentage of people living in communal establishments	Area (hectares)[3]	2001 Density (number of people per hectare)	Number of students away from home[4]
	All people	Males	Females	All people	Males	Females						
a	b	c	d	e	f	g	h	i	j	k	l	m

Notes: 1. 1991 population figures are not available for all areas. For further information see Notes to Tables.
2. Care must be taken when interpreting intercensal population change, as there have been changes in definition between 1991 and 2001, and the 2001 counts have been adjusted to account for under-enumeration.
3. Area measurements are derived from the constituent Output Areas. For further information see Notes to Tables.
4. 'Number of students away from home' is the number of students and schoolchildren in full-time education who would reside in the area were they not living away from home in term-time. Data for the number of students away from home was not available from the 1991 Census.

Table **KS02**

Age[1] structure

All people

Postcode sectors in England and Wales

Area	All people	Percentage of people aged:																	Mean age[2] of population in the area	Median age[2] of population in the area
		0 – 4	5 – 7	8 – 9	10 – 14	15	16 – 17	18 – 19	20 – 24	25 – 29	30 – 44	45 – 59	60 – 64	65 – 74	75 – 84	85 – 89	90 and over		
a	b	c	d	e	f	g	h	i	j	k	l	m	n	o	p	q	r	s	t

Notes: 1. 'Age' is age last birthday.
2. 'Mean age' and 'Median age' are calculated using the underlying single year counts.

Table **KS03**

Living arrangements[1]

All people aged 16 and over in households

Postcode sectors in England and Wales

	All people aged 16 and over in households	Percentage of people aged 16 and over in households						
		Living in a couple		Not living in a couple				
Area		Married or re-married	Cohabiting	Single (never married)	Married or re-married[2]	Separated (but still legally married)	Divorced	Widowed
a	b	c	d	e	f	g	h	i

Notes: 1. The living arrangements variable is different to marital status. It combines information from both marital status and the relationship matrix. Therefore a person living as part of a 'cohabiting couple' could in fact be married (to someone else) but will not appear as married or separated in this classification.
2. A person not living in a couple can be classified married (or re-married) if they denote their marital status as married (or re-married) but have no spouse or partner resident in the household.

Table **KS04**

Marital status

All people aged 16 and over

Postcode sectors in England and Wales

	All people aged 16 and over	Percentage of people aged 16 and over					
Area		Single (never married)	Married	Re-married	Separated (but still legally married)	Divorced	Widowed
a	b	c	d	e	f	g	h

Table **KS05**

Country of birth

All people

Postcode sectors in England and Wales

	All people	Percentage of people born in:						
Area		England	Scotland	Wales	Northern Ireland	Republic of Ireland	Other EU countries[1]	Elsewhere
a	b	c	d	e	f	g	h	i

Notes: 1. 'Other EU countries' includes United Kingdom, part not specified, Ireland, part not specified, Channel Islands and Isle of Man.
2. The European Union as defined on Census day (29 April 2001).

Table KS06

Ethnic group

All people

Postcode sectors in England and Wales

	All people	White			Mixed				Asian or Asian British				Black or Black British			Chinese or other ethnic group	
		British	Irish	Other White	White and Black Caribbean	White and Black African	White and Asian	Other Mixed	Indian	Pakistani	Bangla-deshi	Other Asian	Caribbean	African	Other Black	Chinese	Other ethnic group
								Percentage of people in ethnic groups									
Area	b	c	d	e	f	g	h	i	j	k	l	m	n	o	p	q	r
a																	

Table KS06

Ethnic group

All people

Postcode sectors in Wales

	All people	White			Mixed				Asian or Asian British				Black or Black British			Chinese or other ethnic group		Percentage of all people identifying themselves as Welsh[1]
		British	Irish	Other White	White and Black Caribbean	White and Black African	White and Asian	Other Mixed	Indian	Pakistani	Bangladeshi	Other Asian	Caribbean	African	Other Black	Chinese	Other ethnic group	
								Percentage of people in ethnic groups										
Area	b	c	d	e	f	g	h	i	j	k	l	m	n	o	p	q	r	s
a																		

Note: 1. People identifying themselves as Welsh will appear in both a specific standard ethnic group and in the Welsh group.

Table **KS07**

Religion

All people

Postcode sectors in England and Wales

Area	All people	Christian	Buddhist	Hindu	Jewish	Muslim	Sikh	Other religions	No religion	Religion not stated
		Percentage of people stating religion as:								
a	b	c	d	e	f	g	h	i	j	k

Table **KS08**

Health and provision of unpaid care

All people

Postcode sectors in England and Wales

Area	All people	Limiting long-term illness[1]		General health[3]			Provision of unpaid care[4]			
		Percentage of people with limiting long-term illness[1]	Percentage of people of working age population with limiting long-term illness[2]	Percentage of people whose health was:			All people who provide unpaid care	Percentage of people who provide unpaid care[4]		
				Good	Fairly good	Not good		1 – 19 hours a week	20 – 49 hours a week	50 or more hours a week
a	b	c	d	e	f	g	h	i	j	k

Notes: 1. Limiting long-term illness covers any long-term illness, health problem or disability which limits daily activities or work.
2. Working age population is 16 – 64 years inclusive for men and 16 – 59 years inclusive for women.
3. General health refers to health over the 12 months prior to Census day (29 April 2001).
4. Provision of unpaid care: looking after, giving help or support to family members, friends, neighbours or others because of long-term physical or mental ill-health or disability or problems relating to old age.

Table KS09a

Economic activity

All people aged 16 – 74

Postcode sectors in England and Wales

	ALL PEOPLE AGED 16 – 74	PERCENTAGE OF PEOPLE AGED 16 – 74											PERCENTAGE OF UNEMPLOYED PEOPLE AGED 16 – 74			
		Economically active					Economically inactive								
		Employees								Looking after home/ family						
Area		Part-time[1]	Full-time[1]	Self-employed	Un-employed	Full-time student	Retired	Student		Permanently sick/ disabled	Other	Aged 16 – 24	Aged 50 and over	Who have never worked	Who are long-term unemployed[2]
a	b	c	d	e	f	g	h	i	j	k	l	m	n	o	p

Table KS09b

Economic activity

All males aged 16 – 74

Postcode sectors in England and Wales

	ALL MALES AGED 16 – 74	PERCENTAGE OF MALES AGED 16 – 74											PERCENTAGE OF UNEMPLOYED MALES AGED 16 – 74			
		Economically active					Economically inactive								
		Employees								Looking after home/ family						
Area		Part-time[1]	Full-time[1]	Self-employed	Un-employed	Full-time student	Retired	Student		Permanently sick/ disabled	Other	Aged 16 – 24	Aged 50 and over	Who have never worked	Who are long-term unemployed[2]
a	b	c	d	e	f	g	h	i	j	k	l	m	n	o	p

Table KS09c

Economic activity

All females aged 16 – 74

Postcode sectors in England and Wales

	ALL FEMALES AGED 16 – 74	PERCENTAGE OF FEMALES AGED 16 – 74											PERCENTAGE OF UNEMPLOYED FEMALES AGED 16 – 74			
		Economically active					Economically inactive								
		Employees								Looking after home/ family						
Area		Part-time[1]	Full-time[1]	Self-employed	Un-employed	Full-time student	Retired	Student		Permanently sick/ disabled	Other	Aged 16 – 24	Aged 50 and over	Who have never worked	Who are long-term unemployed[2]
a	b	c	d	e	f	g	h	i	j	k	l	m	n	o	p

Notes: 1. For the Census, part-time is defined as working 30 hours or less a week. Full-time is defined as working 31 or more hours a week.
2. 'Long-term unemployed' are those who stated that they have not worked since 1999 or earlier.

Table **KS10**

Hours worked¹

All people aged 16 – 74 in employment

Postcode sectors in England and Wales

Area	All males aged 16 – 74 in employment	Percentage of males aged 16 – 74 in employment working (hours a week)						All females aged 16 – 74 in employment	Percentage of females aged 16 – 74 in employment working (hours a week)						Average (mean) weekly hours worked:	
		Part-time			Full-time				Part-time			Full-time				
		1 – 5	6 – 15	16 – 30	31 – 37	38 – 48	49 or more		1 – 5	6 – 15	16 – 30	31 – 37	38 – 48	49 or more	Male	Female
a	b	c	d	e	f	g	h	i	j	k	l	m	n	o	p	q

Note: 1. Hours worked is the average number of hours per week worked for the last four weeks before the Census (29 April 2001).

Table KS11a

Industry of employment

All people aged 16 – 74 in employment

Postcode sectors in England and Wales

Area	All people aged 16 – 74 in employment	Agriculture, hunting and forestry	Fishing	Mining and quarrying	Manu-facturing	Electricity, gas and water supply	Construction	Wholesale and retail trade, repair of motor vehicles	Hotels and catering	Transport, storage and commun-ication	Financial inter-mediation	Real estate, renting and business activities	Public admin-istration and defence	Education	Health and social work	Other[1]
						Percentage of people aged 16 – 74 in employment working in:										
a	b	c	d	e	f	g	h	i	j	k	l	m	n	o	p	q

Table KS11b

Industry of employment

All males aged 16 – 74 in employment

Postcode sectors in England and Wales

Area	All males aged 16 – 74 in employment	Agriculture, hunting and forestry	Fishing	Mining and quarrying	Manu-facturing	Electricity, gas and water supply	Construction	Wholesale and retail trade, repair of motor vehicles	Hotels and catering	Transport, storage and commun-ication	Financial inter-mediation	Real estate, renting and business activities	Public admin-istration and defence	Education	Health and social work	Other[1]
						Percentage of males aged 16 – 74 in employment working in:										
a	b	c	d	e	f	g	h	i	j	k	l	m	n	o	p	q

Table KS11c

Industry of employment

All females aged 16 – 74 in employment

Postcode sectors in England and Wales

Area	All females aged 16 – 74 in employment	Agriculture, hunting and forestry	Fishing	Mining and quarrying	Manu-facturing	Electricity, gas and water supply	Construction	Wholesale and retail trade, repair of motor vehicles	Hotels and catering	Transport, storage and commun-ication	Financial inter-mediation	Real estate, renting and business activities	Public admin-istration and defence	Education	Health and social work	Other[1]
						Percentage of females aged 16 – 74 in employment working in:										
a	b	c	d	e	f	g	h	i	j	k	l	m	n	o	p	q

Note: 1. 'Other' includes other community, social and personal service activities, private households with employed persons and extra-territorial organisations and bodies.

Table **KS12a**

Occupation groups

All people aged 16 – 74 in employment **Postcode sectors in England and Wales**

| | | Percentage of people aged 16 – 74 in employment working as: | | | | | | | | |
Area	All people aged 16 – 74 in employment	Managers and senior officials	Professional occupations	Associate professional and technical occupations	Admin-istrative and secretarial occupations	Skilled trades occupations	Personal service occupations	Sales and customer service occupations	Process, plant and machine operatives	Elementary occupations
a	b	c	d	e	f	g	h	i	j	k

Table **KS12b**

Occupation groups

All males aged 16 – 74 in employment **Postcode sectors in England and Wales**

| | | Percentage of males aged 16 – 74 in employment working as: | | | | | | | | |
Area	All males aged 16 – 74 in employment	Managers and senior officials	Professional occupations	Associate professional and technical occupations	Admin-istrative and secretarial occupations	Skilled trades occupations	Personal service occupations	Sales and customer service occupations	Process, plant and machine operatives	Elementary occupations
a	b	c	d	e	f	g	h	i	j	k

Table **KS12c**

Occupation groups

All females aged 16 – 74 in employment **Postcode sectors in England and Wales**

| | | Percentage of females aged 16 – 74 in employment working as: | | | | | | | | |
Area	All females aged 16 – 74 in employment	Managers and senior officials	Professional occupations	Associate professional and technical occupations	Admin-istrative and secretarial occupations	Skilled trades occupations	Personal service occupations	Sales and customer service occupations	Process, plant and machine operatives	Elementary occupations
a	b	c	d	e	f	g	h	i	j	k

Table **KS13**

Qualifications and students

All people aged 16 – 74

Postcode sectors in England and Wales

Area	All people aged 16 – 74	Percentage of people aged 16 – 74 with:						Total number of full-time students and schoolchildren		Percentage of full-time students aged 18 – 74:		
		No qualifications	Highest qualification attained level 1[1]	Highest qualification attained level 2[2]	Highest qualification attained level 3[3]	Highest qualification attained level 4/5[4]	Other qualifications/ level unknown	Aged 16 – 17	Aged 18 – 74	Economically active:		Economically inactive
										In employment	Unemployed	
a	b	c	d	e	f	g	h	i	j	k	l	m

Notes: 1. 1+ 'O' level passes, 1+ CSE/GCSE any grades, NVQ level 1, Foundation GNVQ.
2. 5+ 'O' level passes, 5+ GCSEs (grades A-C), School Certificate, 1+ 'A' levels/'AS' levels, NVQ level 2, Intermediate GNVQ.
3. 2+ 'A' levels, 4+ 'AS' levels, Higher School Certificate, NVQ level 3, Advanced GNVQ.
4. First degree, Higher degree, NVQ levels 4 and 5, HNC, HND, Qualified Teacher Status, Qualified Medical Doctor, Qualified Dentist, Qualified Nurse, Midwife, Heath Visitor.

Table KS14a

National Statistics Socio-economic Classification

All people aged 16 – 74

Postcode sectors in England and Wales

Area	All people aged 16 – 74	Percentage of people aged 16 – 74											
		Large employers and higher managerial occupations	Higher professional occupations	Lower managerial and professional occupations	Intermediate occupations	Small employers and own account workers	Lower supervisory and technical occupations	Semi-routine occupations	Routine occupations	Never worked	Long-term unemployed[1]	Full-time students[2]	Not classifiable for other reasons[3]
a	b	c	d	e	f	g	h	i	j	k	l	m	n

Table KS14b

National Statistics Socio-economic Classification

All males aged 16 – 74

Postcode sectors in England and Wales

Area	All males aged 16 – 74	Percentage of males aged 16 – 74											
		Large employers and higher managerial occupations	Higher professional occupations	Lower managerial and professional occupations	Intermediate occupations	Small employers and own account workers	Lower supervisory and technical occupations	Semi-routine occupations	Routine occupations	Never worked	Long-term unemployed[1]	Full-time students[2]	Not classifiable for other reasons[3]
a	b	c	d	e	f	g	h	i	j	k	l	m	n

Table KS14c

National Statistics Socio-economic Classification

All females aged 16 – 74

Postcode sectors in England and Wales

Area	All females aged 16 – 74	Percentage of females aged 16 – 74											
		Large employers and higher managerial occupations	Higher professional occupations	Lower managerial and professional occupations	Intermediate occupations	Small employers and own account workers	Lower supervisory and technical occupations	Semi-routine occupations	Routine occupations	Never worked	Long-term unemployed[1]	Full-time students[2]	Not classifiable for other reasons[3]
a	b	c	d	e	f	g	h	i	j	k	l	m	n

Notes: 1. For long-term unemployed year last worked is 1999 or earlier.
2. In the NS-SeC classification, all full-time students are recorded in the 'full-time students' category regardless of whether they are economically active or not.
3. 'Not classifiable for other reasons' includes people whose occupation has not been coded.

Table KS15

Travel to work

All people aged 16 – 74 in employment Postcode sectors in England and Wales

Area	All people aged 16 – 74 in employment	Percentage of people who work mainly at or from home	Percentage of people aged 16 – 74 in employment who usually travel to work by:										Average distance (km) travelled[1] to fixed place of work[2]	Percentage of public transport users in households:[3]	
			Underground, metro, light rail, tram	Train	Bus, minibus or coach	Motorcycle, scooter or moped	Driving a car or van	Passenger in a car or van	Taxi or minicab	Bicycle	On foot	Other		With car or van[1]	Without car or van[4]
a	b	c	d	e	f	g	h	i	j	k	l	m	n	o	p

Notes:
1. The distance travelled is a calculation of the straight line between the postcode of place of residence and postcode of workplace.
2. Excludes working at home, no fixed place of work, working at offshore installation, working outside UK.
3. For the purposes of this table, public transport is defined as Underground, metro, light rail or tram; train; bus, minibus or coach.
4. The number of people who travel to work by public transport who live in a household with/without a car or van is expressed as a percentage of the number of people who travel to work by public transport. These figures may not sum to 100% as residents of communal establishments who travel to work by public transport are not included in these counts.

Table **KS16**

Household spaces and accommodation type

All household spaces **Postcode sectors in England and Wales**

	All household spaces			Percentage of all household spaces which are of accommodation type:						
		With no residents		Whole house or bungalow:			Flat, maisonette or apartment:			
Area	With residents	Vacant	Second residence/ holiday accomm- odation	Detached	Semi- detached	Terraced (including end-terrace)	Purpose- built block of flats or tenement	Part of a converted or shared house (including bed-sits)	In commercial building	Caravan or other mobile or temporary structure
a	b	c	d	e	f	g	h	i	j	k

Table **KS17**

Cars or vans[1]

All households **Postcode sectors in England and Wales**

		Percentage of households (number of cars or vans)					All cars or vans in the area[2]
Area	All households	None	One	Two	Three	Four or more	
a	b	c	d	e	f	g	h

Notes: 1. Includes any company car or van if available for private use.
2. 'All cars or vans in the area' includes only those cars and vans owned by, or available for use by, households. This count is not exact as households with more than 10 cars or vans are counted as having 10 cars or vans.

Table **KS18**

Tenure

All households **Postcode sectors in England and Wales**

		Percentage of households:						
		Owner occupied			Rented from:			
Area	All households	Owns outright	Owns with a mortgage or loan	Shared ownership[1]	Council (local authority)	Housing Association / Registered Social Landlord[2]	Private landlord or letting agency	Other[3]
a	b	c	d	e	f	g	h	i

Notes: 1. Pays part rent and part mortgage.
2. Includes Housing Co-operative and Charitable Trust.
3. Includes employer of a household member and relative or friend of a household member and living rent free.

Table **KS19**

Rooms, amenities, central heating and lowest floor level

All households

Postcode sectors in England and Wales

Area	All households	Average household size	Average number of rooms per household	Percentage of households:					Lowest floor level			
				With an occupancy rating of -1 or less[1]	With central heating and sole use of bath/shower and toilet	Without central heating or sole use of bath/shower and toilet	With central heating, with sole use of bath/shower and toilet	With central heating, without sole use of bath/shower and toilet	Basement or semi-basement	Ground level (street level)	1st/2nd/3rd or 4th floor	5th floor or higher
a	b	c	d	e	f	g	h	i	j	k	l	m

Note: 1. The occupancy rating provides a measure of under-occupancy and overcrowding. For example a value of -1 implies that there is one room too few and that there is overcrowding in the household. The occupancy rating assumes that every household, including one person households, requires a minimum of two common rooms (excluding bathrooms).

Table **KS20**

Household composition

All households

Postcode sectors in England and Wales

Area	ALL HOUSE-HOLDS	PERCENTAGE OF HOUSEHOLDS COMPRISING:															
		One family and no others												Other households			
		One person		All pensioners	Married couple households			Cohabiting couple households			Lone parent households						
		Pensioner	Other		No children	With dependent children[1]	All children non-dependent	No children	With dependent children[1]	All children non-dependent	With dependent children[1]	All children non-dependent	With dependent children[1]	All student	All pensioner	Other
a	b	c	d	e	f	g	h	i	j	k	l	m	n	o	p	q

Note: 1. A dependent child is a person in a household aged 0 – 15 (whether or not in a family) or a person aged 16 – 18 who is a full-time student in a family with parent(s).

Table **KS21**

Households with limiting long-term illness and dependent children

All households

Postcode sectors in England and Wales

		Percentage of households				
		No adults in employment		With dependent children[1]		With one or more person with a limiting long-term illness
Area	All households	With dependent children[1]	Without dependent children[1]	All ages	Aged 0 – 4	
a	b	c	d	e	f	g

Note: 1. A dependent child is a person in a household aged 0 – 15 (whether or not in a family) or a person aged 16 – 18 who is a full-time student in a family with parent(s).

Table **KS22**

Lone parent households with dependent children

All lone parent households with dependent children

Postcode sectors in England and Wales

	Male lone parent[3]			Female lone parent[3]		
All lone parent households with dependent children[1]	Total	Percentage in part-time[2] employment	Percentage in full-time[2] employment	Total	Percentage in part-time[2] employment	Percentage in full-time[2] employment
b	c	d	e	f	g	h

Notes: 1. A dependent child is a person in a household aged 0 – 15 (whether or not in a family) or a person aged 16 – 18 who is a full-time student in a family with parent(s).
2. For the Census, part-time is defined as working 30 hours or less a week. Full-time is defined as working 31 or more hours a week.
3. For the purposes of this table, a lone parent is defined as a parent with a dependent child living in a household with no other parents (whether related to that dependent child or not). This definition is to be distinguished from the standard definition of lone parent used in other tables.

Table KS23

Communal establishment residents[1]

All communal establishment residents

Postcode sectors in England and Wales

	ALL COMMUNAL ESTABLISHMENTS	NUMBER OF RESIDENTS	PERCENTAGE OF RESIDENTS LIVING IN:										PERCENTAGE OF RESIDENTS IN COMMUNAL ESTABLISHMENTS WITH A LIMITING LONG-TERM ILLNESS	
			Medical and care establishments									Other establishments	Medical and care establishments	Other establishments
			NHS		Local authority		Housing association	Other						
Area			Psychiatric	Other	Children's home	Other		Nursing homes	Residential care homes	Children's homes	Other			
a	b	c	d	e	f	g	h	i	j	k	l	m	n[2]	o[2]

Notes:
1. 'Residents' excludes staff and families of staff.
2. The number of residents in medical and care establishments who have a limiting long-term illness is expressed as a percentage of the number of residents in medical and care establishments. The number of residents in other establishments who have a limiting long-term illness is expressed as a percentage of the number of residents in other establishments.

Table KS24

Migration

All People

Postcode sectors in England and Wales

	All People	Percentage of these who are migrants	Percentage of all people:					All people in ethnic groups other than 'White'	Percentage of these who are migrants	Percentage of people in ethnic groups other than 'White':				
			Who moved into the area		No usual address 1 year ago	Who moved within the area	Who moved out of the area[1]			Who moved into the area		No usual address 1 year ago	Who moved within the area	Who moved out of the area[1]
Area			From within the UK	From outside the UK						From within the UK	From outside the UK			
a	b	c	d	e	f	g	h	i	j	k	l	m	n	o

Note: 1. People whose address one year ago was within the area but whose current address is outside the area but within the UK, expressed as a percentage of current population of the area. People in these columns do not appear in columns relating to total migrants.

Table **KS25**

Knowledge of Welsh

All people aged 3 and over　　　　　　　　　　　　　　　　　　　　　　**Postcode sectors in Wales**

| | | Percentage of people aged 3 and over | | | | | |
Area	All people aged 3 and over	Understands spoken Welsh only[1]	Speaks but does not read or write Welsh	Speaks and reads but does not write Welsh	Speaks, reads and writes Welsh	Other combination of skills	No knowledge of Welsh
a	b	c	d	e	f	g	h

Note:　1. 'Understands spoken Welsh only' means the person understands spoken Welsh but has no other skills in the language.

PCP01

People resident in area

All people resident in area	Postcode sectors in England and Wales	
	All people resident in area	Percentage of all people resident in area

TOTAL POPULATION RESIDENT

Sex
  Male
  Female

Age
  0 – 15
  16 – 24
  25 – 34
  35 – 44
  45 – 54
  55 – 64
  65 and over

Ethnic group
  White
    British
    Irish
    Other White

  Mixed
    White and Black Caribbean
    White and Black African
    White and Asian
    Other mixed

  Asian or Asian British
    Indian
    Pakistani
    Bangladeshi
    Other Asian

  Black or Black British
    Black Caribbean
    Black African
    Other Black

  Chinese or other ethnic group
    Chinese
    Other ethnic group

PCP02

People resident in area aged 16 – 74

All people aged 16 – 74	Postcode sectors in England and Wales	
	All people aged 16 – 74	Percentage of all people aged 16 – 74

Economic activity
All people aged 16 – 74

Economically active
  Employed
  Unemployed
  Full time student

Economically inactive
  Retired
  Student
  Looking after home/family
  Permanently sick/disabled
  Other

PCP03

Households resident in area

All households	Postcode sectors in England and Wales	
	All households	Percentage of all households resident in the area

TOTAL HOUSEHOLDS RESIDENT

Households in dwelling type
Detached
Semi-detached
Terraced
Flat
Other

Number of rooms
1 or 2
3, 4, 5 or 6
7 or more

Tenure
Owner occupied
Rented from private landlord
Rented from Council or Housing Association
Other

Household size
1 person
2
3
4
5 or more

Car ownership
None
1
2
3+

NS-SeC of Household Reference Person (HRP)
1. Higher managerial and professional occupations
    1.1 Large employers and higher managerial occupations
    1.2 Higher professional occupations
2. Lower managerial and professional occupations

3. Intermediate occupations
4. Small employers and own account workers
5. Lower supervisory and technical occupations
6. Semi-routine occupations
7. Routine occupations

8. Never worked and long-term unemployed
    L14.1 Never worked
    L14.2 Long-term unemployed
Not classified
    L15 Full-time students
    L17 Not classifiable for other reasons

Households with dependent children

Notes: *1. NS-SeC stands for National Statistics Socio-economic Classification.*
        2. In the NS-SeC classification all full-time students are recorded in the 'full-time students'
        category regardless of whether they are economically active or not.
        3. 'Not classifiable for other reasons' includes people whose occupation has not been coded
        and those who cannot be allocated to a NS-SeC category.